Contents

1. Night Lights   4
by David Hill

2. Sounds in Space   15
by John O'Brien

3. Return Ticket, Please!   18
by David Hill

4. Space Junk   26
by Jill Brasell

Glossary   31

Index   32

Night Lights

by David Hill

What looks like a 15-mile-long potato that's been hit by something traveling at 100 miles a second?

What's pizza-colored and has volcanoes all over it?

What's covered with ice, is hidden by orange clouds, and is seven light years away from Earth?

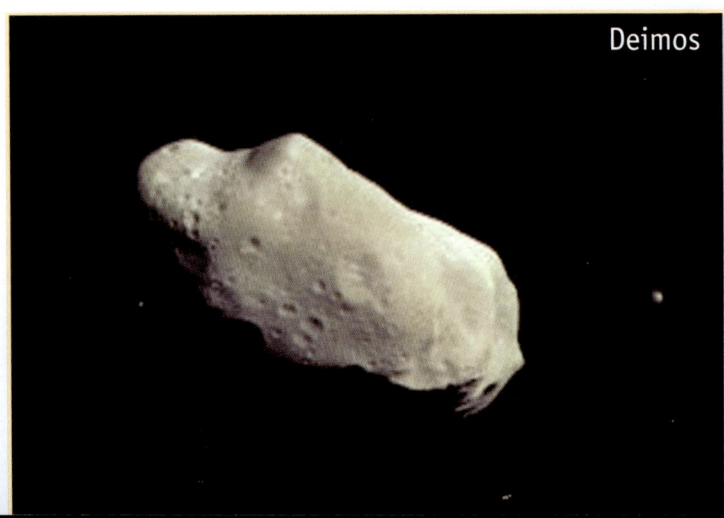

Deimos

Night Lights

A Cruise around
the Solar System

Learning Media

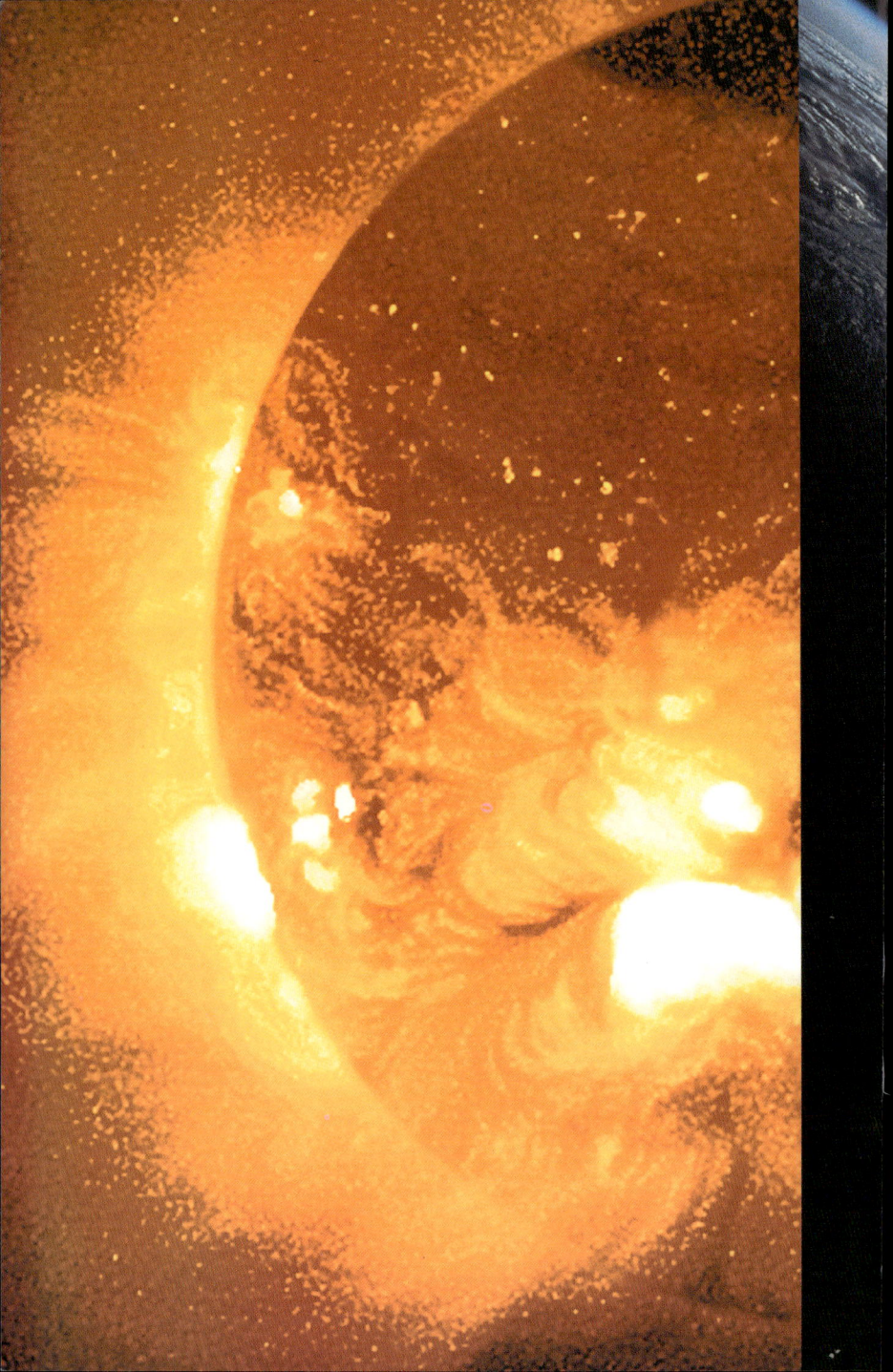

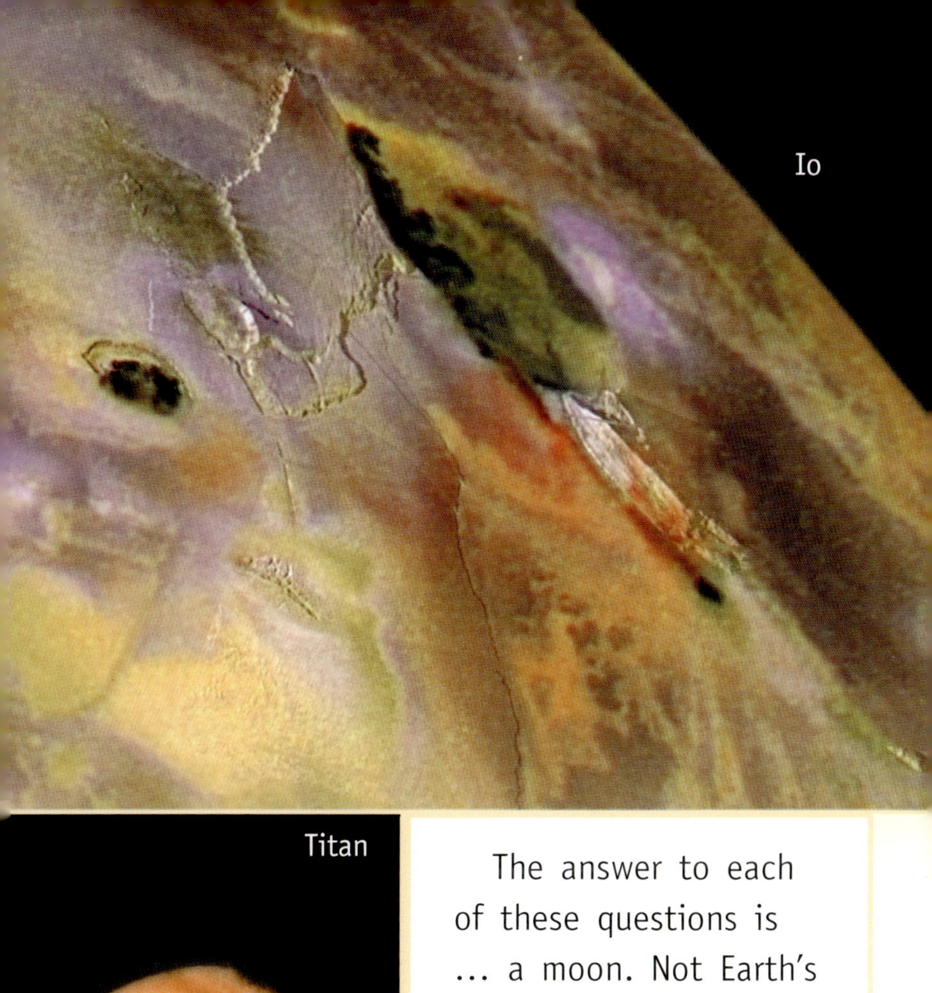

Io

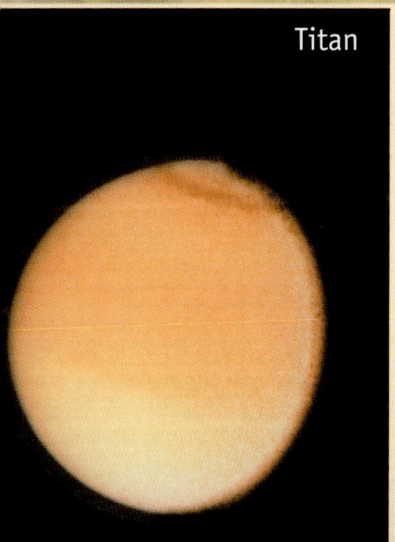

Titan

The answer to each of these questions is ... a moon. Not Earth's Moon but the moons belonging to Mars, Jupiter, and Saturn. (Deimos, Io, and Titan, to be exact.)

There are over eighty moons **orbiting** the nine planets of our **solar system**. **Astronomers** have been able to study most of them, using spacecraft and telescopes. Most moons orbit the four biggest planets – Jupiter has twenty, Saturn twenty-nine, Uranus eighteen, and Neptune eleven. Earth, Mars, and Pluto have only one or two each. Mercury and Venus have no moons at all. That's because they are so close to the Sun that its powerful **gravity** would drag anything smaller than a planet toward it.

Saturn and some of its moons

So, what's the difference between a planet and a moon?

- *A planet orbits a star (like our Sun).*
- *A moon orbits a planet.*
- *A moon is always smaller than its own planet, but some moons are bigger than some other planets.*

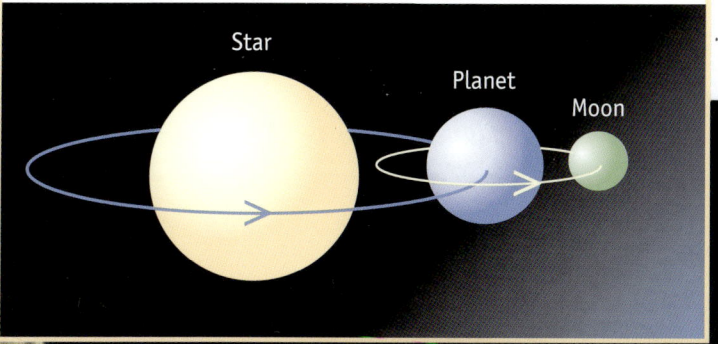

Planet-Moon Fact File

- *Mars's two moons, Phobos and Deimos, are each only about the size of a small island.*
- *There are seven moons (including Earth's) that are bigger than the smallest planet of our solar system, Pluto.*
- *There are two moons that are bigger than Mercury.*
- *Pluto's moon, Charon, is half as big as Pluto.*
- *Two of Jupiter's moons are big enough to be seen from Earth through binoculars – they appear as bright dots on either side of the planet.*
- *Some of Jupiter's closest moons take only 7 hours to orbit the giant planet, but its more distant moons take 2 years.*
- *The surface of Jupiter's moon Io is pulled in and out by Jupiter's enormous gravity. Because of this, Io's crust heats up and erupts in volcanoes.*

Eruption on Io

Names of the Moons

The names of moons usually match the names of their planets. Mars, named after the Roman god of war, has moons named after the gods of fear and panic, Phobos and Deimos. Neptune was the sea god, and this planet's moons are named after sea creatures and sea monsters. The moons of Jupiter are named after that god's many human partners. Our Moon was called Luna by the ancient Romans.

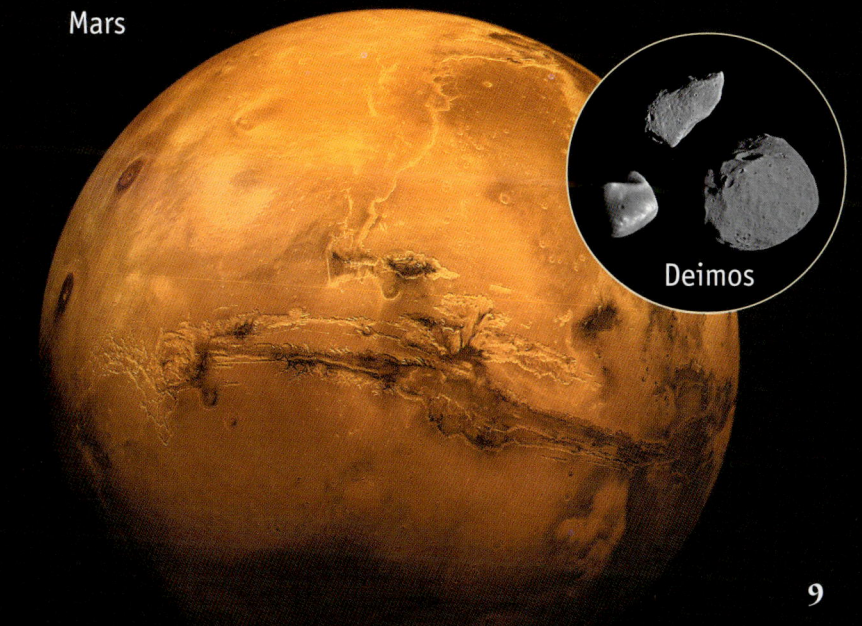

Mars

Deimos

Our Moon

Earth's moon is 239,000 miles away. It doesn't produce any light of its own but just reflects sunlight. We can't see all of the lit-up part from Earth all the time – that's why the Moon seems to change shape. It takes about a month to "grow" from a new Moon to a full Moon and then "shrink" back again. These different shapes are called the phases of the Moon.

Phases of the Moon

New   Crescent   Quarter   Full   Quarter   Crescent   New

Astronomers have different ideas about where Earth's Moon came from. One is that Earth's gravity captured the Moon as it drifted past. Another is that the Moon began as a blob of **molten** rock that was flung off by Earth. The latest theory is that a Mars-sized planet smashed into Earth billions of years ago. The huge impact threw vast amounts of rock and dust into space, where they clumped together and formed the Moon.

Earth's Moon is the fifth largest moon in the solar system. It's a quarter of the size of Earth but weighs a lot less because it's far less **dense**. There's no air or water on the Moon. So, without any wind or rain to disturb them, astronauts' footprints in the moon dust could last forever. Well, almost – the Moon is moving away from Earth by about half a mile a year. In a few million years, it will escape from Earth's orbit and plunge into the Sun.

The dark patches you can see on the Moon are called "seas." They're actually plains where volcanic lava once flowed. You can also see hundreds of craters where **meteorites** have slammed into the Moon's surface. The biggest craters are about 190 miles across. There are craters all over the Moon, but we can't see the ones on the far side because that side is always turned away from us.

Moon Mysteries

- *Why are there **geysers** on Neptune's biggest moon, Triton?*

- *Were Pluto and its moon Charon once parts of a bigger planet, or were they both moons of Uranus that were knocked out of orbit by a huge **meteor**?*

- *What caused the enormous crater that covers nearly half of Saturn's moon Mimas?*

- *Why does the surface of Uranus's moon Miranda have lines that look as if they've been made by a huge rake?*

- *Could there be life on the warm slopes of Io's volcanoes?*

SOUNDS IN SPACE

by John O'Brien

In movies, space is a very noisy place. Rockets roar along; their blasters and lasers scream and whine. Things explode with a noise like thunder. On TV and in cartoons, it's just the same. Starships rumble from planet to planet. Noise erupts everywhere.

But that's not how it is. Space is never noisy, and there's a very good reason why.

Sounds are caused when something **vibrates**. If you look at a guitar string after it has been plucked, you can see it vibrating. Sometimes you can feel sound vibrations.

When something vibrates, the air around it begins to vibrate. The vibrations spread out through the air, like ripples spreading out across a pool of water. When they reach your ears, you hear them as a sound.

Vibrations can travel through solid things as well as air. They can travel through water too. But they cannot travel through nothing.

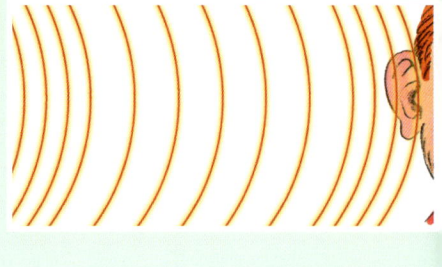

In space, there is no air. There is nothing for vibrations to travel through. So sounds *cannot* travel through space.

If you were in space and a passing rocket fired off all its weapons, you would hear nothing. If something exploded in the distance, you would hear nothing. If a giant starship rushed past at half the speed of light, you would hear nothing.

A comet colliding into Jupiter

The movies and television and cartoons have got it wrong. Space is not a noisy place at all! Space is a place of SILENCE!

Return Ticket, Please!

by David Hill

The first person ever to stand on the Moon was Neil Armstrong on 20 July 1969. As he stepped out of his spacecraft, he said, "That's one small step for a man ... one giant leap for mankind."

What would *you* say if you were the first person ever to step onto another planet?

The United States, working with other countries, is planning a new mission to Mars by the year 2020. And this time, there'll be *people* on board.

Spacecraft without crews have been landing on Mars since the 1970s. In 1997, the spacecraft *Pathfinder* landed on Mars, and a small, six-wheeled robot called *Sojourner* trundled around, testing rocks and measuring the wind speed.

A trip to Mars takes a long time. The astronauts who walked on the Moon took three days to get there, traveling at almost 4,000 miles per hour in their *Apollo* spacecraft. At the same speed, it would take them twenty *months* to reach Mars. Even with today's more powerful rockets, it would still take at least seven months to reach Mars.

Why Mars?

You might think that Venus would be the first planet that people would plan to visit. After all, it's 18.5 million miles closer to Earth than Mars is. But the surface temperature on Venus is 900°F – hot enough to melt lead. The clouds drizzle acid. Winds tear overhead at 250 miles per hour, and the **air pressure** is enough to crush a human body. Nobody seems very keen to take a walk on Venus!

Mars is the only planet in our solar system (apart from Earth) where people might be able to survive.

Visitors to Mars probably won't find any life there as there doesn't seem to be any water. But scientists think that primitive life may have existed there in the past. But even primitive life could not exist without water, so maybe there was water there once. Future spacecraft may check whether there is still any water beneath the dry surface of the planet.

More Than a Year in Space?

Astronauts for the trip to Mars will have to be chosen very carefully. Carrying all the supplies for such a long journey means that there'll be very little room on board.

How would you like to spend fourteen months (there and back) shut up in a small room with the same three or four people? What if you didn't like each another? Imagine if someone snored!

It might be possible to grow some fresh vegetables on the way. (One problem is that in **zero gravity**, the plants' roots won't know which way to grow!) All moisture, including washing water and sweat, will have to be collected and recycled for drinking water.

It's a Long Way Away

Radio messages travel at the speed of light. But Mars is so far away that it will take messages four minutes and twenty seconds to get back to Earth. If an astronaut asked mission control, "Help! Where's the fire extinguisher?", it would be nearly nine minutes before a reply came back.

Spacecraft have millions of working parts. If a spacecraft develops engine trouble halfway to Mars, the astronauts can't call for a tow truck. Everything will have to work perfectly, all the time.

Problems on Mars

If people do finally arrive on Mars, there will be a few problems to solve.

There's no **oxygen** on Mars, only **carbon dioxide**. So the astronauts could not breathe the air.

Mars is also a very cold planet. Even in summer, the temperature seldom reaches 40°F. At night, it drops to minus 220°F.

They might not be able to see very much, either. On Mars, dust storms can last for months, filling the sky with pink dust.

A trip to Mars hardly sounds like the ideal vacation! But people have always wanted to discover new places. And who wants to sit around at home all the time?

So – return ticket to Mars, anyone?

SPACE JUNK

by Jill Brasell

There's not much space left in the "space" (called Near Earth Orbit or NEO) that lies between Earth's atmosphere and outer space. Thousands, perhaps millions, of objects are whizzing around up there, including hundreds of working satellites and thousands of dead ones, rocket boosters, and other bits of broken and cast-off hardware. The amount of space junk doubles about every seven years.

This illustration shows how thick the cloud of debris orbiting Earth is.

Space junk is a serious danger to space missions and the **telecommunications** instruments that are orbiting Earth. All that debris is hurtling through space at speeds of up to 25,000 miles per hour. At this rate, colliding with a piece of metal the size of a small marble would be like being under a 400-pound safe that had been dropped from 100 feet up.

So if you're planning a trip to outer space, here's a few of the things you might need to dodge on your way through NEO:

- *Old satellites*
- *Rocket boosters*
- *Fragments of space shuttle windows (63 have had to be replaced since 1981.)*
- *The silver glove that astronaut Ed White dropped during his first spacewalk*
- *Caskets containing the ashes of rich, dead space-nuts who wanted their remains launched into orbit*
- *Natural space junk, such as meteors*
- *Thousands of tiny metal fragments resulting from explosive space-junk collisions.*

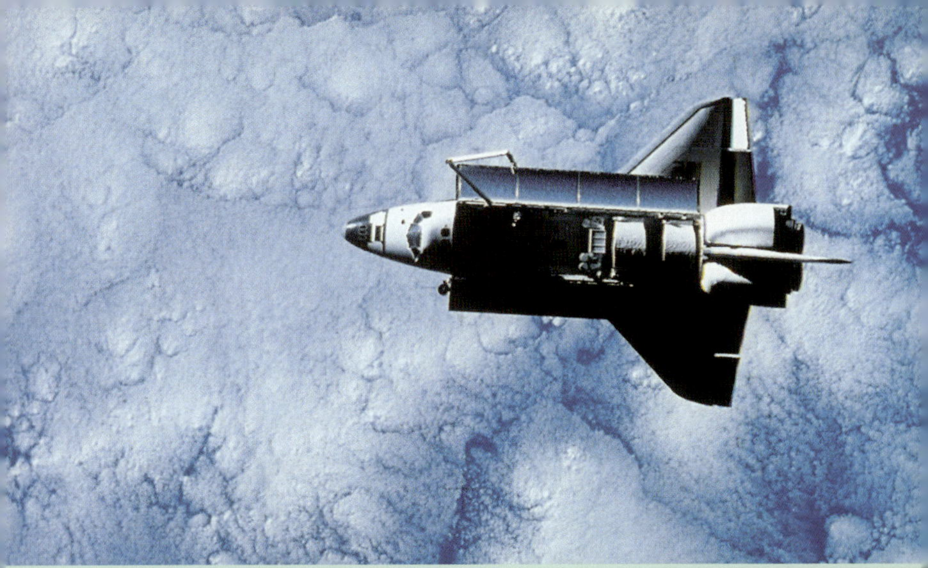

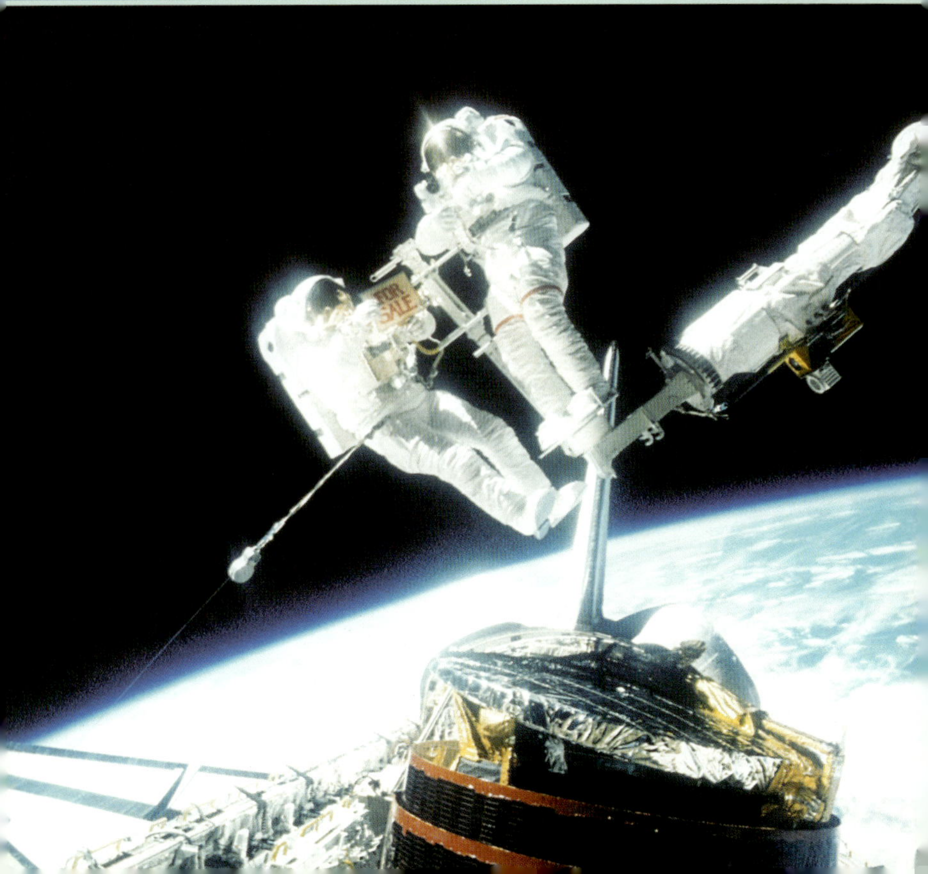
FOR SALE

Glossary

(These words are printed in bold type the first time they appear in the text.)

air pressure: the weight of the atmosphere pressing down

astronomers: scientists who study space

bacteria: very tiny, simple life forms

carbon dioxide: a gas that people and animals breathe out but can't live in

dense: heavy for its size

geysers: natural fountains of hot water

gravity: the force that pulls a small object toward a much larger object

meteor: a stray piece of rock flying through space

meteorite: the remains of a meteor after it has landed on a planet

molten: melted by extreme heat

orbiting: circling endlessly around something

oxygen: the gas that must be in the air for animals and people to survive

solar system: a star (such as our Sun) and the planets that orbit around it

telecommunications: methods of communicating using telephone technology

vibrates: makes small, very rapid movements back and forth

zero gravity: a state that exists in space, where nothing falls and there is no "down" or "up"

Index

Mars –
    life on   20–21, 24–25
    visits to   19, 22, 25
Moon, the –
    origin   11
    phases   10
    seas   13
moons –   7
    Jupiter   5, 6, 8
    Mars   4, 5, 6, 8
    Mercury   6, 8
    names of   9
    Neptune   6, 14
    Pluto   6, 8, 14
    Saturn   5, 6, 14
    Uranus   6, 14
planets   7
space –
    junk   26–29
    sounds   15–17
star   7
Sun, the   6, 7
Venus   20